Earth Astronomy and Astrology Guide for Beginners

Renée and Roggie Clark

Explore The Whole Astronomy and Astrology Guide for Beginners Series

Discover the fascinating worlds of astronomy and astrology through the **Astronomy and Astrology Guide for Beginners Series**, a collection of easy-to-understand guides that explore the planets through both scientific knowledge and centuries of human interpretation.

Each book focuses on a single planet and explores it from two distinct perspectives. Learn about the planet's physical characteristics, orbit, atmosphere, moons, exploration, discoveries, and place within our solar system, while also discovering its history, mythology, symbolism, zodiac associations, and traditional role in astrology.

These beginner-friendly guides are designed to help readers understand not only what we scientifically know about the planets, but also how different cultures and astrological traditions have interpreted their meanings throughout history.

You'll also learn how to observe planets in the night sky when possible, recognize important astronomical events, understand basic planetary movements, and explore the traditional meanings associated with planets in birth charts, zodiac signs, houses, aspects, and transits.

Series Titles Include

- Mercury Astronomy and Astrology Guide for Beginners
- Venus Astronomy and Astrology Guide for Beginners
- Earth Astronomy and Astrology Guide for Beginners
- Mars Astronomy and Astrology Guide for Beginners
- Jupiter Astronomy and Astrology Guide for Beginners
- Saturn Astronomy and Astrology Guide for Beginners
- Uranus Astronomy and Astrology Guide for Beginners
- Neptune Astronomy and Astrology Guide for Beginners
- And More

Whether you're interested in astronomy, stargazing, planetary science, astrology, zodiac signs, birth charts, mythology, or simply curious about the worlds that share our solar system, the **Astronomy and Astrology Guide for Beginners Series** provides a welcoming introduction to each planet and the stories, discoveries, and traditions surrounding it.

From the physical worlds studied by astronomers to the symbolic meanings developed by astrologers and ancient cultures, each guide offers a broader way to explore humanity's enduring fascination with the planets.

Readers can also find additional books, printable resources, and educational materials by searching "**Roggie Clark Books**" online.

Explore Our Crystal Guides for Beginners Series

Discover the fascinating world of crystals through the **Crystal Guide for Beginners Series**, a collection of easy-to-understand guides designed for readers who want to learn about gemstones, crystal traditions, and metaphysical practices.

Each book focuses on a single crystal and explores its history, symbolism, traditional healing associations, zodiac connections, rituals, journaling prompts, and practical everyday uses.

You'll also learn how to identify, care for, and work with your crystal while gaining a deeper understanding of the beliefs and traditions that have surrounded it for generations.

Series Titles Include

- Amethyst Guide for Beginners
- Rose Quartz Guide for Beginners
- Clear Quartz Guide for Beginners
- Citrine Guide for Beginners
- Black Tourmaline Guide for Beginners
- Moonstone Guide for Beginners
- Labradorite Guide for Beginners
- Selenite Guide for Beginners
- Tiger's Eye Guide for Beginners
- Fluorite Guide for Beginners
- And More

Whether you're interested in crystal collecting, spirituality, meditation, journaling, astrology, or simply learning something new, the **Crystal Guide for Beginners Series** provides a welcoming introduction to some of the world's most popular crystals.

Readers can also find additional books, printable resources, and educational materials by searching "Roggie Clark Books" online.

Continue Learning At Our Website:

Roggieclarkbooks.com

Continue exploring educational resources, tools, books, workbooks, and learning materials designed to help readers continue building knowledge and practical skills. Visit: www.roggieclarkbooks.com

Inside you'll find:

- Downloadable resources
- Educational worksheets
- Beginner-friendly guides
- Printable activities
- Companion materials for books and courses

A growing library of educational content for readers, collectors, learners, parents, teachers, and lifelong students.

Copyright

Title: Earth Astronomy and Astrology Guide for Beginners
Authors: Renée and Roggie Clark

Cover Design: Roggie Clark
Editing: Renée Clark
Printed in the United States of America

Disclaimer

This book is for entertainment purposes only and is not intended to substitute for professional medical or mental health advice, diagnosis, or treatment. Always seek the advice of your physician or qualified mental health provider with any questions you may have regarding a medical or mental health condition.

Astronomy and Astrology Disclaimer

The information contained in this book is provided for educational, informational, historical, and entertainment purposes only.

The discussions of astronomy, celestial objects, planetary movements, constellations, the zodiac, astrology, birth charts, planetary associations, astrological houses, aspects, transits, lunar cycles, spiritual interpretations, and related traditions are intended to introduce readers to these subjects in an accessible manner.

Astronomical information presented in this book is based on established scientific knowledge and general educational sources. However, this book is intended as a beginner's guide and should not be considered a substitute for academic instruction, professional scientific research, or specialized astronomical resources. Scientific understanding, classifications, measurements, and discoveries may change as new research and observations become available.

Astrological concepts discussed in this book are based on historical traditions, cultural beliefs, symbolic systems, and contemporary astrological practices. Astrology and its interpretations should not be understood as scientifically established methods for predicting future events, determining personality, diagnosing conditions, or guaranteeing particular outcomes. Interpretations of zodiac signs, planets, houses, aspects, birth charts, transits, and other astrological concepts may vary among traditions and practitioners.

Any astrological readings, interpretations, personality descriptions, compatibility discussions, forecasts, or spiritual associations presented in this book are intended for educational, reflective, and entertainment purposes. They should not be treated as definitive statements about an individual's personality, relationships, future, health, finances, career, or life circumstances.

This book does not provide medical, psychological, psychiatric, therapeutic, legal, financial, investment, relationship, or other professional advice. Readers should consult an appropriately qualified professional when making decisions involving health, mental health, finances, legal matters, or other circumstances requiring professional guidance.

Readers should not delay seeking professional advice, disregard professional recommendations, or discontinue prescribed treatment based on astrological interpretations or other information presented in this book.

Important Solar Observation Safety Warning

Never look directly at the Sun with the naked eye, through binoculars, through a telescope, through a camera lens, or through any other optical or magnifying device unless you are using appropriate equipment and solar filters specifically designed and properly installed for safe solar observation. Looking at the Sun improperly can cause serious and permanent eye injury.

Ordinary sunglasses, homemade filters, tinted glass, photographic film, and other improvised materials should not be relied upon for safe solar viewing. Readers should follow the safety instructions provided by reputable astronomical organizations and the manufacturers of any solar observation equipment being used.

Children should perform solar observation activities only with appropriate adult supervision and properly designed solar viewing equipment.

Readers are responsible for using telescopes, binoculars, cameras, solar viewers, and other observational equipment safely and according to the manufacturer's instructions. The author and publisher are not responsible for injuries or damages resulting from improper equipment use or unsafe observation practices.

Any decisions made or actions taken based on the information presented in this book are solely the responsibility of the reader. The author and publisher disclaim liability for any loss, injury, claim, damage, or other consequence arising from the use, interpretation, or misuse of the information contained herein.

By reading this book, you acknowledge that astronomy and astrology are approached differently within this series: astronomy is presented as a scientific study of celestial objects and phenomena, while astrology is presented as a historical, cultural, symbolic, and interpretive tradition. Individual interpretations and experiences with astrology may vary.

Table of Contents

Introduction

Earth is the only known planet to support life, making it unique within our solar system and, so far, throughout the observable universe. It is our home, the world that has shaped every civilization, ecosystem, and human experience throughout history. While we often think of Earth as familiar, it is an extraordinary planet filled with remarkable geological processes, diverse climates, abundant water, and an atmosphere perfectly suited to sustaining life. Beyond its scientific significance, Earth has also been viewed as sacred across many cultures, inspiring countless spiritual traditions, creation stories, and symbolic interpretations. This guide explores both the scientific understanding of Earth and the historical, cultural, and traditional symbolic meanings associated with our home planet while clearly distinguishing scientific evidence from cultural and spiritual traditions.

What This Guide Covers

This guide provides a beginner-friendly introduction to Earth from both scientific and traditional perspectives. You'll learn how Earth formed approximately 4.5 billion years ago, what makes it capable of supporting life, and how its atmosphere, oceans, continents, and internal structure work together to create a dynamic and ever-changing world. You'll also explore Earth's place in mythology, cultural traditions, and spiritual beliefs, along with the symbolic meanings that have been associated with nature, balance, stability, and life itself. Throughout this guide, astronomy is presented as an evidence-based scientific discipline, while spiritual and symbolic traditions are explored as important parts of human history and culture. Together, these perspectives offer a well-rounded appreciation of the remarkable planet we call home.

A Beginner's Introduction to Earth

Earth is the third planet from the Sun and the largest of the four **terrestrial planets** in the inner solar system. Approximately seventy-one percent of its surface is covered by oceans, while the remaining land consists of continents, islands, mountains, deserts, forests, grasslands, and countless other landscapes. Earth is currently the only known planet with stable liquid water on its surface, an oxygen-rich atmosphere, and the conditions necessary to support an incredible diversity of life.

Earth's interior is made up of several layers, including a solid inner core, a liquid outer core, a thick mantle, and a relatively thin crust. The movement of these layers drives **plate tectonics**, creating earthquakes, volcanoes, mountain ranges, and the continual reshaping of the planet's surface. Earth's magnetic field also protects the planet from much of the harmful solar radiation that constantly streams outward from the Sun.

Although Earth may seem ordinary because we live here, scientists recognize it as one of the most remarkable planets ever discovered. Its combination of liquid water, moderate temperatures, protective atmosphere, active geology, and thriving ecosystems makes it unlike any other known world.

Understanding Astronomy and Astrology

Although Earth is central to astronomy, its role in astrology differs from that of the other planets. **Astronomy** studies Earth as one of the planets orbiting the Sun, examining its formation, structure, atmosphere, climate, geology, and relationship with the rest of the solar system. Scientists also use Earth as the standard by which many other planets are measured, making it an essential reference point in planetary science.

In traditional astrology, Earth itself is generally not treated as a planet because astrology is historically based on the viewpoint of an observer standing on Earth's surface. Instead, Earth serves as the reference point from which the movements of the Sun, Moon, and planets are interpreted. Even so, many modern spiritual traditions associate Earth with qualities such as grounding, stability, abundance, nourishment, growth, and connection with nature.

Throughout this guide, these different perspectives remain clearly distinguished. Scientific explanations describe Earth's physical characteristics and place within the universe, while symbolic traditions explore the meanings that people have historically associated with the natural world and humanity's relationship with it.

Last Thoughts

Earth is both our home and one of the greatest scientific wonders known to humanity. Its unique ability to support life, constantly changing geology, and breathtaking diversity make it an extraordinary planet worthy of continual exploration and appreciation. As you continue through this guide, you'll discover how Earth formed, why it became capable of sustaining life, and how it has inspired scientific discovery, mythology, cultural traditions, and spiritual reflection throughout human history. Together, these perspectives reveal why understanding Earth is one of the most meaningful journeys we can undertake.

Chapter 1: What Is Earth?

Earth is our home planet and the only world currently known to support life. While it may seem ordinary because we experience it every day, Earth is an extraordinary planet with a unique combination of characteristics that make it unlike any other world yet discovered. Its abundant liquid water, breathable atmosphere, active geology, protective magnetic field, and stable climate have allowed countless forms of life to evolve over billions of years. Understanding what Earth is and how it formed provides the foundation for appreciating its remarkable place within our solar system and the wider universe.

Physical Characteristics

Earth is the third planet from the Sun and the largest of the four inner rocky planets. It has an average diameter of approximately 7,900 miles (12,742 kilometers), making it significantly larger than Mercury and Mars but smaller than the giant outer planets. Its nearly spherical shape is caused by gravity pulling matter evenly toward its center while its rotation creates a slight bulge around the equator.

Unlike many other planets, Earth has vast oceans that cover about seventy-one percent of its surface. The remaining land includes continents, islands, mountain ranges, forests, grasslands, deserts, wetlands, and frozen polar regions. This incredible diversity of environments supports millions of different species, from microscopic organisms to enormous mammals and towering trees.

Earth's atmosphere is another defining characteristic. It consists primarily of nitrogen and oxygen, with smaller amounts of argon, carbon dioxide, water vapor, and other gases. This atmosphere protects life by filtering much of the Sun's harmful ultraviolet radiation, regulating temperature, and providing the oxygen needed by most living organisms.

Beneath the surface, Earth is composed of several layers. The crust forms the thin outer shell where we live, while the mantle beneath it slowly moves over millions of years. At the center lies a liquid outer core surrounding a solid inner core made primarily of iron and nickel. The movement of molten metals within the outer core generates Earth's magnetic field, which helps shield the planet from charged particles emitted by the Sun.

Earth also has one natural satellite—the Moon. The Moon influences ocean tides, helps stabilize Earth's axial tilt, and has played an important role in shaping both the planet's environment and human history.

Formation

Scientists estimate that Earth formed approximately 4.54 billion years ago during the early formation of the solar system. At that time, a massive cloud of gas and dust known as the **solar nebula** began collapsing under its own gravity. As the young Sun formed at the center, the remaining material gradually came together through countless collisions, eventually creating the planets.

Earth began as a hot, molten world. Frequent impacts from asteroids and other space debris released enormous amounts of energy, keeping much of the planet's surface in a molten state. During this early period, heavier elements such as iron and nickel sank toward the center, while lighter materials rose toward the surface. This process created the layered structure that still exists today.

As Earth gradually cooled, its first solid crust formed. Volcanic activity released gases that helped create the early atmosphere, although it was very different from the oxygen-rich atmosphere we know today. Water vapor eventually condensed as temperatures fell, forming clouds that produced rain over millions of years. This rainfall filled low-lying regions, creating Earth's earliest oceans.

Over billions of years, Earth's atmosphere changed dramatically as simple life forms evolved. Tiny photosynthetic organisms began releasing oxygen into the atmosphere, gradually transforming the planet into an environment capable of supporting increasingly complex life. This slow but remarkable process ultimately made Earth the biologically rich world we know today.

Where Found

Earth is located in the inner region of our solar system, orbiting the Sun between Venus and Mars. It follows a nearly circular orbit at an average distance of about 93 million miles (150 million kilometers) from the Sun. This distance is often referred to as one **astronomical unit (AU)** and serves as a standard measurement for distances within the solar system.

Earth occupies a region commonly called the **habitable zone**, where temperatures allow liquid water to exist on the planet's surface under the right atmospheric conditions. While this location is an important factor in Earth's ability to support life, scientists recognize that many additional factors—including the atmosphere, magnetic field, geological activity, and chemical composition—also contribute to the planet's long-term habitability.

Earth completes one orbit around the Sun in approximately 365.25 days, creating our calendar year. At the same time, it rotates once every twenty-four hours, producing the familiar cycle of day and night. Its axis is tilted by about 23.5 degrees, causing the changing seasons experienced across much of the world.

Although Earth is only one of eight planets in our solar system, it serves as the reference point for much of planetary science. Researchers compare newly discovered planets to Earth when evaluating their size, composition, atmosphere, and potential to support life. Even as thousands of planets have been discovered orbiting distant stars, Earth remains the only world where life has been confirmed.

Last Thoughts

Earth is far more than the place where we live—it is an exceptionally complex and dynamic planet that continues to surprise scientists with new discoveries. From its formation billions of years ago to its oceans, atmosphere, magnetic field, and thriving ecosystems, every part of Earth contributes to the conditions that make life possible. Understanding Earth's physical characteristics, how it formed, and where it is located within the solar system provides an essential foundation for exploring its history, symbolism, natural processes, and unique role in both science and human culture throughout the rest of this guide.

Chapter 2: History and Symbolism

Earth has always been more than just the ground beneath our feet. Long before people understood geology, plate tectonics, or the formation of planets, civilizations around the world looked to the land, mountains, rivers, forests, and skies to explain their place in the universe. Earth was viewed as the source of life, a provider of food and shelter, and the foundation upon which all living things depended. Although cultures developed different stories and beliefs about Earth's origins, many shared a deep respect for the natural world and recognized humanity's close relationship with it. Today, scientific discoveries explain Earth's physical history, while historical and symbolic traditions offer valuable insight into how people have understood and appreciated their home throughout the ages.

Historical Uses

Throughout history, Earth has served as humanity's home, workplace, classroom, and source of survival. Early civilizations depended entirely on the land for food, water, building materials, and natural resources. Agriculture allowed communities to settle permanently, leading to the development of villages, cities, trade networks, and eventually great civilizations.

Many ancient societies carefully observed the changing seasons and natural cycles to determine when to plant and harvest crops. These observations led to the creation of calendars that helped communities organize farming, religious celebrations, and daily life. The rhythms of nature became deeply connected to human survival, encouraging people to pay close attention to the changing environment around them.

Earth's natural resources have also shaped human history. Stone, clay, metals, timber, fresh water, and fertile soil enabled technological advances and economic development. Entire civilizations rose near rivers, coastlines, and fertile plains where the land could support growing populations. As societies advanced, they developed new ways to mine minerals, cultivate crops, construct buildings, and harness Earth's resources for transportation, manufacturing, and energy production.

Even today, Earth's geology continues to influence where people live and how societies develop. Mountain ranges, rivers, coastlines, deserts, forests, and climate zones all affect transportation, agriculture, industry, and settlement patterns across the globe.

Spiritual and Cultural Traditions

Across cultures and throughout history, Earth has often been viewed as a symbol of life, nourishment, stability, and connection. Many Indigenous cultures emphasize living in harmony with nature, recognizing that humans are part of Earth's larger ecosystem rather than separate from it. Respect for the land, animals, plants, and natural cycles remains central to many of these traditions.

In Greek mythology, **Gaia** was regarded as the primordial personification of Earth and the mother of many gods and living beings. She represented fertility, creation, and the foundation upon which all life depended. Similar Earth-centered figures appear in many other cultures, each reflecting the importance of the land as both a physical and spiritual source of existence.

Numerous cultures have also developed creation stories explaining how Earth came into being. While these accounts differ widely in their details and are distinct from scientific explanations, they reveal humanity's enduring desire to understand where the world came from and humanity's place within it. These stories often emphasize themes of balance, responsibility, renewal, and the interconnectedness of life.

Seasonal celebrations have also played an important role throughout history. Festivals marking spring, harvest, solstices, and equinoxes reflected people's close relationship with Earth's natural rhythms. These traditions honored the changing seasons and expressed gratitude for the resources that sustained communities year after year.

Traditional Symbolic Meanings

In many symbolic traditions, Earth represents stability, security, patience, endurance, and practical wisdom. Because it provides the foundation for all terrestrial life, it has long been associated with reliability, consistency, and growth through steady effort rather than sudden change.

Earth is also commonly viewed as a symbol of grounding. In many modern wellness and spiritual practices, grounding refers to cultivating a sense of balance, presence, and connection with the physical world. Activities such as spending time in nature, gardening, hiking, or simply walking outdoors are often described as ways to reconnect with the natural environment and promote a greater sense of well-being. While experiences vary from person to person, these practices reflect humanity's long-standing appreciation for the restorative qualities of nature.

Another traditional symbolic meaning associated with Earth is abundance. Fertile soil produces crops, forests provide resources, rivers supply water, and ecosystems support countless forms of life. As a result, Earth has often been viewed as representing prosperity, generosity, and the ongoing cycles of renewal found throughout nature.

Earth also symbolizes responsibility. Unlike many symbolic traditions centered on personal qualities, Earth's symbolism frequently emphasizes stewardship and care. Many cultures encourage people to recognize that future generations depend upon how today's societies manage natural resources, protect ecosystems, and preserve the planet's biodiversity.

Although these symbolic meanings are not scientific descriptions of the planet itself, they continue to influence literature, philosophy, art, environmental movements, and personal reflection around the world.

Last Thoughts

Earth has shaped every civilization, culture, and generation throughout human history. It has provided the resources necessary for survival, inspired countless myths and traditions, and become a lasting symbol of life, stability, and connection with nature. While science explains Earth's physical origins and ongoing evolution, historical and symbolic traditions help us understand the many ways people have valued and interpreted their relationship with the natural world. Together, these perspectives deepen our appreciation for the extraordinary planet that continues to sustain and inspire humanity.

Chapter 3: Metaphysical and Healing Properties

For thousands of years, people have viewed Earth as more than a physical planet. Across many spiritual traditions, philosophies, and modern wellness practices, Earth has symbolized stability, nourishment, resilience, and connection with the natural world. These interpretations are rooted in cultural and spiritual beliefs rather than scientific evidence, but they continue to influence meditation practices, personal reflection, mindfulness, and nature-based traditions around the world. Understanding these symbolic associations offers insight into why Earth has remained such a powerful source of inspiration throughout human history.

Grounding and Stability

One of the most widely recognized symbolic qualities associated with Earth is grounding. In many spiritual and wellness traditions, grounding refers to developing a sense of balance, presence, and connection with one's surroundings. Rather than becoming overwhelmed by constant distractions or emotional turmoil, grounding encourages individuals to remain focused on the present moment and the realities immediately before them.

Earth's steady nature has made it a universal symbol of reliability and consistency. Mountains, forests, rivers, and landscapes often represent permanence in contrast to the constantly changing circumstances of everyday life. Because the planet has sustained life for billions of years through countless natural cycles, it has become a reminder that growth often occurs gradually through patience and persistence.

Many people incorporate time in nature into their personal wellness routines. Walking through parks, hiking, gardening, or simply sitting outdoors are commonly practiced ways of slowing down and reconnecting with the natural environment. While scientific research supports certain mental and physical benefits of spending time in nature, the symbolic idea of becoming "grounded" remains a personal and cultural interpretation rather than a measurable scientific phenomenon.

Earth's symbolism also encourages resilience. Just as landscapes continually change through erosion, volcanic activity, seasonal cycles, and weather, people often view challenges as opportunities for gradual growth rather than permanent setbacks.

Abundance and Nurturing Energy

Earth has long been associated with abundance because it provides the resources necessary for life. Fertile soil produces crops, forests offer shelter and materials, rivers provide fresh water, and ecosystems support an incredible diversity of living organisms. As a result, many symbolic traditions connect Earth with nourishment, generosity, and the continual renewal found throughout nature.

In spiritual traditions, abundance is not always limited to financial prosperity. It may also refer to gratitude, meaningful relationships, personal growth, creativity, health, knowledge, or opportunities. Earth symbolizes the idea that growth often begins with caring for what has already been planted, whether that means developing new skills, strengthening relationships, or nurturing personal goals over time.

Earth is also frequently associated with patience. Seeds require time to grow into plants, forests develop over centuries, and landscapes are shaped across millions of years. These natural processes remind many people that meaningful progress rarely happens instantly and that lasting results are often built through consistent effort.

Because Earth continually supports countless forms of life, it has also become a symbol of compassion, caregiving, and responsibility. Many traditions encourage gratitude for the natural resources that sustain humanity while emphasizing the importance of protecting them for future generations.

Balance and Connection

Another important symbolic meaning associated with Earth is balance. Nature operates through interconnected systems in which plants, animals, water, air, soil, and climate all influence one another. Observing these relationships has inspired many philosophies that encourage people to seek balance within their own lives.

Earth reminds us that no living organism exists entirely alone. Every species depends upon countless interactions within larger ecosystems. This understanding has influenced many cultural traditions that emphasize cooperation, respect for nature, and awareness of humanity's relationship with the environment.

Many meditation and mindfulness practices use imagery connected with Earth to encourage feelings of calm, stability, and centeredness. Visualizing deep roots extending into the ground, imagining standing firmly upon solid earth, or focusing on the steady rhythms of nature are common symbolic exercises used to cultivate emotional balance and presence. These practices are intended for personal reflection and should not be viewed as scientific or medical treatments.

Earth also symbolizes interconnectedness across generations. The landscapes we inherit today have been shaped by billions of years of geological processes and thousands of years of human history. Likewise, the choices made today influence the world that future generations will experience. This perspective encourages thoughtful stewardship, personal responsibility, and appreciation for the shared home that all people inhabit.

Last Thoughts

Earth's traditional metaphysical associations reflect humanity's enduring relationship with the natural world. Across cultures and generations, it has symbolized grounding, stability, abundance, patience, resilience, balance, and our deep connection to life itself. While these meanings arise from spiritual traditions and personal beliefs rather than scientific research, they continue to provide many people with meaningful opportunities for reflection and personal growth. Whether viewed through the lens of science or symbolism, Earth serves as a powerful reminder of the importance of caring for ourselves, one another, and the extraordinary planet we all call home.

Chapter 4: Spiritual Correspondences

Throughout history, Earth has represented far more than the physical planet beneath our feet. In many spiritual traditions, philosophical systems, and modern wellness practices, Earth symbolizes stability, nourishment, endurance, and our connection to the natural world. While these correspondences are based on cultural and spiritual beliefs rather than scientific evidence, they continue to influence meditation, symbolic interpretation, and personal reflection. Understanding these traditional associations provides another perspective on why Earth has remained such an important symbol across civilizations.

Chakras

In many modern spiritual traditions that incorporate the **chakra system**, Earth is most commonly associated with the **Root Chakra**. Located at the base of the spine, the Root Chakra is traditionally connected with feelings of safety, security, stability, and one's connection to the physical world.

Earth's dependable nature makes it a fitting symbol for these qualities. Just as the planet provides a stable foundation for life, the Root Chakra is traditionally viewed as the energetic foundation upon which the rest of a person's well-being is built. When individuals feel grounded and secure, they may find it easier to navigate life's challenges with confidence and resilience.

Practices commonly associated with strengthening this symbolic connection include spending time outdoors, walking in natural settings, gardening, mindful breathing, and developing consistent daily routines. These activities are intended as personal wellness or spiritual practices and should not be considered medical or scientific treatments.

Some traditions also associate Earth with physical well-being because it represents the body, the natural environment, and the practical aspects of everyday life. This symbolism encourages people to care for both themselves and the world around them.

Zodiac Associations

Unlike the other planets, Earth does not traditionally rule any zodiac sign because astrology developed from the perspective of observers standing on Earth's surface. Instead, Earth serves as the central reference point from which the movements of the Sun, Moon, and planets are interpreted.

Even so, many modern astrologers and spiritual writers symbolically connect Earth with the **Earth signs**: Taurus, Virgo, and Capricorn. These signs are traditionally associated with qualities such as practicality, dependability, patience, responsibility, persistence, and a strong appreciation for the physical world.

Taurus is often connected with comfort, beauty, nature, and material stability. Virgo is traditionally associated with organization, service, careful attention to detail, and continual improvement. Capricorn commonly symbolizes discipline, long-term planning, perseverance, and steady achievement.

These symbolic relationships reflect characteristics traditionally associated with the Earth element rather than with the planet Earth itself. As with all astrological interpretations, these associations are part of a longstanding symbolic tradition and are not supported by scientific evidence.

Planetary Symbolism

Earth's symbolic meaning extends beyond any single spiritual tradition. Across cultures, it has consistently represented life, home, nourishment, responsibility, endurance, and the interconnectedness of all living things.

Earth reminds many people that growth often requires patience. Just as forests mature over decades, mountains are shaped over millions of years, and ecosystems develop through countless natural cycles, meaningful personal growth is often viewed as a gradual process rather than an immediate transformation.

Another enduring symbolic theme is stewardship. Because Earth supports every known form of life, many philosophical and spiritual traditions encourage respect for nature and responsible care for the environment. This symbolism emphasizes that humans are not separate from nature but are participants within a much larger living system.

Earth also represents balance between giving and receiving. The planet continually provides resources that sustain life, while humanity bears the responsibility of protecting those resources for future generations. This reciprocal relationship has inspired countless teachings about gratitude, conservation, compassion, and mindful living.

Perhaps most importantly, Earth symbolizes belonging. Every person, regardless of culture, nationality, or belief system, shares the same planetary home. This simple reality has inspired many traditions to emphasize unity, cooperation, and recognition of humanity's shared connection to the natural world.

Last Thoughts

Earth's spiritual correspondences highlight humanity's enduring desire to understand its relationship with the world around us. Whether symbolizing grounding through the Root Chakra, reflecting the qualities traditionally associated with the Earth signs, or representing life, balance, stewardship, and belonging, Earth continues to inspire meaningful reflection across many cultures and traditions. Although these correspondences arise from symbolic and spiritual beliefs rather than scientific research, they provide another lens through which many people appreciate the remarkable planet we all share.

Chapter 5: Observing and Appreciating Earth

Unlike every other planet discussed in this series, Earth is not something we observe from afar—it is the world we experience every moment of our lives. Because we live on its surface, it can be easy to overlook just how extraordinary our planet truly is. Earth offers an ever-changing landscape of mountains, oceans, forests, deserts, glaciers, rivers, grasslands, and countless ecosystems that make it one of the most diverse worlds known. Appreciating Earth involves more than learning scientific facts; it also means developing a deeper awareness of the natural processes, environments, and living systems that surround us every day.

Observing Earth's Natural Wonders

One of the best ways to appreciate Earth is simply by observing the natural world. Every landscape tells part of the planet's long geological story. Mountain ranges reveal the movement of tectonic plates, river valleys show the power of erosion, volcanoes remind us that Earth's interior remains active, and coastlines constantly change through the forces of wind, waves, and weather.

The sky also provides countless opportunities for observation. Watching the sunrise and sunset, observing cloud formations, following the changing phases of the Moon, or identifying planets and constellations helps us better understand Earth's place within the larger solar system. Seasonal changes, shifting daylight hours, and changing weather patterns all demonstrate the dynamic relationship between Earth and the Sun.

Many people also develop a greater appreciation for Earth by visiting national parks, nature preserves, botanical gardens, forests, beaches, mountains, or other protected natural areas. These environments offer opportunities to observe ecosystems that have developed over thousands or even millions of years.

Keeping a nature journal can further deepen this appreciation. Recording seasonal changes, observing local wildlife, identifying plants, or simply documenting time spent outdoors encourages greater awareness of the remarkable diversity that exists within even familiar surroundings.

Appreciating Earth's Role in Supporting Life

Earth is currently the only known planet capable of supporting such an extraordinary diversity of life. From microscopic bacteria living beneath the ocean floor to giant trees reaching toward the sky, every ecosystem depends upon a delicate balance of environmental conditions that has developed over billions of years.

Earth's atmosphere provides breathable air and helps regulate temperatures. Its oceans influence weather, store heat, and support vast marine ecosystems. Freshwater rivers and lakes sustain countless species, while healthy soils allow plants to grow and provide food for animals and people alike. Every component of Earth's environment contributes to a larger interconnected system that supports life across the planet.

Recognizing these connections often leads to a deeper appreciation of the natural world. A single tree contributes oxygen, provides habitat for wildlife, stabilizes soil, and participates in Earth's water and carbon cycles. Likewise, wetlands filter water, coral reefs protect coastlines, and forests help regulate climate while supporting incredible biodiversity.

Understanding these relationships reminds us that even small actions can influence the health of larger ecosystems. Appreciating Earth often begins by recognizing how closely human well-being is connected to the health of the planet itself.

Developing a Lifelong Connection with Earth

Appreciating Earth is not limited to scientific study or outdoor adventures. It can also become a lifelong habit of curiosity, gratitude, and responsible stewardship. Learning about geology, ecology, meteorology, oceanography, astronomy, or environmental science provides countless opportunities to deepen one's understanding of the natural world.

Many people also strengthen their connection with Earth through simple everyday activities. Gardening, hiking, birdwatching, camping, photographing landscapes, caring for local green spaces, or learning about native plants and wildlife all encourage greater awareness of the environment. These experiences often help people notice details that might otherwise go unnoticed.

Earth also teaches patience. Natural processes often unfold gradually over years, centuries, or even millions of years.

Rivers slowly carve canyons, forests regenerate after disturbances, mountains rise and erode, and ecosystems continually adapt to changing conditions. Observing these long-term processes reminds us that meaningful change often occurs through steady progress rather than sudden transformation.

Perhaps most importantly, appreciating Earth encourages a sense of shared responsibility. Every person depends upon the same atmosphere, oceans, freshwater resources, and ecosystems. Caring for these natural systems helps preserve the planet for future generations while allowing countless species—including our own—to continue thriving.

Last Thoughts

Earth is the only home humanity has ever known, and its beauty, complexity, and diversity continue to inspire wonder with every new discovery. From observing magnificent landscapes and changing seasons to understanding the intricate systems that sustain life, there are countless ways to develop a deeper appreciation for our remarkable planet. Whether your interest lies in science, nature, conservation, or simply spending more time outdoors, taking time to observe and appreciate Earth fosters a greater understanding of the world around us and reminds us of the incredible privilege of living on such an extraordinary planet.

Chapter 6: Bringing It All Together

Earth is unique among all the planets we have explored in this series because it is not a distant world that we observe through telescopes or visit with spacecraft—it is the planet we call home. Every mountain climbed, every ocean crossed, every forest explored, and every city built exists on this remarkable world. Throughout this guide, you've learned about Earth's scientific characteristics, its place within the solar system, its role in human history and culture, and the symbolic meanings that people have associated with it for thousands of years. Together, these perspectives reveal that Earth is both an extraordinary planet and an endless source of discovery.

Connecting Science and Symbolism

One of the most important lessons in studying Earth is recognizing that science and symbolism answer different kinds of questions. Science explains how Earth formed, how its atmosphere developed, why plate tectonics reshape its surface, and how its oceans, climate, and ecosystems support life. These explanations are built upon observation, experimentation, measurement, and continual scientific investigation.

Symbolic traditions, on the other hand, explore humanity's relationship with Earth through stories, philosophy, spirituality, and culture. Across civilizations, Earth has represented stability, nourishment, patience, responsibility, and life's interconnectedness. These meanings developed through human experience rather than scientific testing and continue to influence art, literature, mindfulness practices, and cultural traditions around the world.

Rather than competing with one another, these perspectives offer different ways of appreciating the same planet. Scientific knowledge helps us understand Earth's physical processes, while symbolic traditions encourage reflection on humanity's place within the natural world. Together, they remind us that Earth can be appreciated through both careful observation and thoughtful contemplation.

Studying both perspectives also encourages respect for different ways people have sought understanding throughout history. While modern science continues to expand our knowledge of Earth's physical nature, historical traditions reveal how generations of people found meaning in the landscapes, seasons, and natural cycles surrounding them.

Continuing Your Journey

Learning about Earth is only the beginning of understanding our place in the universe. Every scientific discovery raises new questions about how planets form, how climates change, and whether other worlds might someday be found that support life. Earth serves as the foundation for much of planetary science because scientists compare newly discovered planets to the one we know best.

If your interest lies in science, consider exploring subjects such as geology, meteorology, oceanography, ecology, astronomy, or environmental science. Visiting museums, nature centers, observatories, or national parks provides opportunities to see many of Earth's natural processes firsthand. Reading about current scientific discoveries and space missions can also deepen your understanding of how Earth fits into the broader story of the solar system.

If you're interested in history, mythology, or symbolic traditions, you may enjoy learning how different cultures understood nature and the changing seasons. Exploring Indigenous knowledge, ancient civilizations, philosophical traditions, and environmental ethics can provide valuable perspectives on humanity's long relationship with the planet.

Above all, remain curious. Curiosity encourages observation, thoughtful questions, and a willingness to continue learning. Whether you're watching a thunderstorm roll across the horizon, examining rocks along a hiking trail, identifying constellations on a clear night, or learning about Earth's ancient history, every experience offers another opportunity to better understand the remarkable world we inhabit.

Why Earth Still Matters Today

Earth remains the most important planet we will ever study because it is currently the only known world that supports life. Every scientific breakthrough related to Earth's climate, oceans, atmosphere, geology, ecosystems, and biodiversity helps us better understand not only our own planet but also the countless worlds being discovered beyond our solar system.

Modern research continues to reveal new information about Earth's complex systems. Scientists monitor changing climates, study earthquakes and volcanoes, explore the deepest parts of the oceans, investigate ancient fossils, and examine how ecosystems respond to environmental change.

These discoveries improve our understanding of Earth's past while helping societies prepare for the future.

Earth also serves as a reminder of our shared responsibility. Regardless of nationality, language, or culture, every person depends upon the same atmosphere, freshwater resources, fertile soils, and interconnected ecosystems. Protecting these natural systems helps preserve the planet's ability to support future generations while maintaining the incredible diversity of life that exists today.

Beyond its scientific importance, Earth continues to inspire artists, writers, philosophers, educators, and spiritual traditions around the world. Its landscapes, seasons, and natural beauty encourage wonder, creativity, and appreciation for the world we often take for granted. Whether viewed through the lens of science or symbolism, Earth remains one of the greatest sources of knowledge and inspiration available to humanity.

Last Thoughts

Earth is far more than the third planet from the Sun—it is a dynamic, living world that continues to shape every aspect of human existence. Its unique combination of water, atmosphere, geology, and biodiversity has made it the only known home for life, while its beauty and complexity have inspired countless cultures, traditions, and discoveries throughout history. By exploring Earth through both scientific understanding and historical symbolism, we gain a deeper appreciation for the planet that sustains us every day. As you continue your journey of learning, may Earth remind you that even the most familiar places can hold extraordinary stories, and that curiosity about our own world is often the first step toward understanding the universe itself.

Last Thoughts

Congratulations on completing your journey through Earth.

Although Earth is the planet we know best, it is also one of the most extraordinary worlds ever discovered. It is a planet where oceans, continents, mountains, deserts, forests, rivers, glaciers, and countless ecosystems exist together in a delicate balance that has supported life for billions of years. Every scientific discovery reveals just how complex and remarkable our home truly is, reminding us that even familiar places still hold countless mysteries waiting to be explored.

Throughout this guide, you've learned how Earth formed during the early history of the solar system, how its atmosphere, oceans, and interior evolved over time, and why its unique combination of conditions has allowed life to flourish. You've also explored the many ways civilizations have understood Earth through mythology, cultural traditions, and symbolic interpretations that emphasize stability, nourishment, responsibility, and humanity's connection with nature. While these symbolic traditions differ from scientific explanations, both reflect humanity's enduring fascination with the world beneath our feet.

One of the greatest lessons Earth teaches is that everything is connected. The atmosphere influences the oceans, the oceans affect the climate, living organisms depend upon healthy ecosystems, and human societies rely on Earth's natural resources every day. Understanding these connections encourages both curiosity and responsibility, reminding us that caring for our planet benefits not only ourselves but also future generations.

Earth also reminds us that remarkable things often go unnoticed because they are familiar. The sunrise that marks the beginning of each day, the changing seasons, the growth of forests, the movement of rivers, the formation of mountains, and the incredible diversity of life are all part of an ongoing story that has unfolded over billions of years. Taking time to appreciate these everyday wonders can deepen our understanding of both the natural world and our place within it.

If this guide has inspired your curiosity, continue exploring. Learn about Earth's geology, weather, oceans, wildlife, ecosystems, and changing climate. Spend time outdoors observing the natural world, visit museums and national parks, follow new scientific discoveries, and continue learning about the remarkable processes that shape our planet. Every new question leads to another opportunity for discovery.

Thank you for joining us on this exploration of Earth. We hope this guide has helped you better understand not only the science behind our remarkable planet but also the history, culture, and symbolic traditions that have surrounded it throughout human civilization. Whether you continue your journey through astronomy, environmental science, geology, history, or simply by spending more time appreciating the natural world around you, may Earth always remind you that the greatest adventures in learning often begin right at home.

Meet the Authors

Renée Clark

An author.
A poet.
A healer.
A lover of language and storytelling.
My Divine purpose is to create stories that heal, inspire, and entertain.
I am thankful for God blessing me with this amazing gift and talent. I am thankful God blessed me with the courage to embrace my talent, and I am thankful to be able to share my art with you.
Sending you love, light, and kindness.
With my deepest appreciation,
Renée

Roggie Clark

Roggie "Raj" Clark is an entrepreneur and a 10-year marketing professional with a proven track record in e-commerce and digital business growth.

Having built and managed successful online stores with thousands of sales and consulted for numerous marketing clients, Roggie has mastered the intricacies of online marketing, branding, and consumer behavior. He has also written several popular books on digital marketing, SEO (search engine optimization), and programming (HTML, CSS, and JavaScript).

As a former marketing manager at a fintech company—a role he secured through his own marketing and financial expertise—Roggie gained firsthand experience in financial strategy and wealth-building.

This journey inspired him to share insights with amateur investors, helping them save money, make smarter financial decisions, and build long-term wealth. Raj currently lives in Maryland with his sister and niece.

Teach Your Kids The Science Behind Gemstones With This Book

Gemstones for Kids is designed to help young readers learn what gemstones are, how they are formed, and why they are special. Using real images and easy to understand explanations, this book makes learning about gemstones both fun and educational.

Check Out A Science Fiction Adventure for Kids About Amethyst, Confidence, and Trusting Yourself

This thoughtful science fiction adventure blends imagination, emotional growth, and real-world gemstone knowledge into a chapter book that helps young readers explore science, problem-solving, and self-trust naturally and engagingly.

A Simple Guide to Buying, Evaluating, and Building Wealth with Natural Gems

The book also breaks down the top gemstones worth considering, along with a simple plan to build your first gemstone portfolio. You will discover how to buy safely, avoid common mistakes, store your gems correctly, and understand when and how to sell.

Learn Macro and Micro Economics, With This One Book

Whether you're a student, a professional, or simply someone eager to understand the world around you, this book is your perfect introduction to the fundamentals of micro and macroeconomics.

https://www.amazon.com/Macro-Micro-Economics-Workbook-Beginners-ebook/dp/B0FVPMFX6B

The Strategic Guide to U.S. Mint Gold and Silver Coin Investing

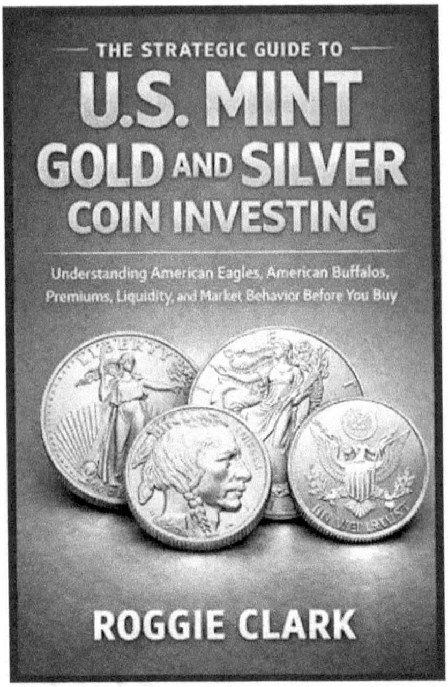

Rather than offering a general introduction to precious metals, this guide examines how American Gold Eagles, American Gold Buffalos, and Silver American Eagles behave in real investor markets.

https://www.amazon.com/Strategic-Guide-U-S-Silver-Investing/dp/B0GQKJF9K6

www.ingramcontent.com/pod-product-compliance
Lightning Source LLC
LaVergne TN
LVHW050749061126
851857LV00013B/1069
* 9 7 9 8 1 8 2 7 3 4 3 5 6 *